NATURAL ASTAXANTHIN
Hawaii's Supernutrient

By
William Sears, M.D.

Illustrations by Debbie Maze

ISBN #: 978-0-9792353-4-4

DR. SEARS' LEADING TITLES

From the Sears Parenting Library

The A. D. D. Book
The Attachment Parenting Book
The Autism Book
The Baby Book
The Baby Sleep Book
The Birth Book
The Breastfeeding Book
The Discipline Book
The Family Nutrition Book
The Fussy Baby Book
The Healthiest Kid in the Neighborhood
The Healthy Pregnancy Book
The N. D. D. Book
The Portable Pediatrician
The Pregnancy Book
The Premature Baby Book
The Successful Child
The Vaccine Book

From the Sears Children's Library

Baby on the Way
Eat Healthy, Feel Great
What Baby Needs
You Can Go to the Potty

Other Sears Books

The Inflammation Solution
The Omega-3 Effect
Prime-Time Health

TABLE OF CONTENTS

Foreword

Though I've spent my medical career with an open mind to new interventions and treatments in the field of health and medicine, the scientist side of me has always been a skeptic – show me the research.

With a dizzying array of new products, claims, and information in the natural products and dietary supplement industry, it's often difficult to discern the wheat from the chaff. Sure, a strong marketing campaign or successful viral video could lead to a brief bump in awareness and sales, but the acceptance and long-term sustainability of an ingredient or product is really dependent upon strong, ongoing scientific support.

When I first heard about astaxanthin *("asta-'zan-thin")*, the scientific research was in its infancy. Since then I have been surprised at the rate at which new discoveries associated with this impressive ingredient have been made. The potency of astaxanthin, which can be achieved with only 4–12 milligrams, sharply contrasts with that of other natural ingredients, especially in the herbal world, where several grams are required to achieve positive research outcomes.

The effectiveness and favorable dosing open up a whole world of opportunities for astaxanthin. Not only can it be used alone, but it can also be used in combination with other ingredients to support the health of the brain, joints, eyes, heart, and immune system, and it even has cosmetic applications.

Besides all the promising data in the research and the favorable dosing amount, the purity, availability, and sustainability of astaxanthin are real bright spots when compared with other ingredients. The growing and cultivation of astaxanthin in Hawaii, a pristine climate, with pure water and raw materials, help create a finished product that is both potent, reliable, and essentially devoid of the pesticide residues and heavy metals that can be problematic in many other ingredients in the industry. It's a relief to know that we all have access to a pure and reliable ingredient.

The future is indeed bright for astaxanthin. It's going to be exciting to see its continued development and progress going forward.

Jason Theodosakis, MD, MS, MPH, FACPM, author of the New York Times #1 Bestselling Book The Arthritis Cure, *and clinical associate professor of Family and Community Medicine at the University of Arizona.*

A Note to My Readers

We all search for the secrets to health. Reading this book will enlighten you about one of nature's top secrets: people who eat the most antioxidants tend to enjoy the best health. You are about to embark on a journey to the land and sea of antioxidants and learn which type is the most healthful.

Among my professional colleagues, I'm known as the show-me-the-science doctor. I won't take, eat, or prescribe any medicine or start a health habit unless I've read the science on it. My family, my patients, and my readers trust that simple authenticity. I'm also known as the science-made-simple doctor. In fact, when I'm paired with a researcher in talks at medical meetings we jokingly call our lecture the scientist-and-simpleton talk. So, readers, I assure you this book is not only science-based, but simple and fun to read. Enjoy!

William Sears, M.D.
Capistrano Beach, CA
January 18, 2015

FINDING THE SECRETS TO HEALTH

Following a life-threatening illness at the age of 57, I searched for the secrets to health – and found them. At the prime of my life, with eight children and then five grandchildren, in addition to a thriving medical practice and publishing career, I had a lot to live for.

Eighteen years later, at the age of 75, I'm enjoying wonderful health with the blood pressure and blood chemistry of a younger man, and I take no regular prescription medicines. I enjoy the simple health goal that everyone wants: Everything works and nothing hurts. Naturally, I was motivated to answer the question: What do people do who live the healthiest and longest? The answer: Go fish! In searching for the top health tips, I found one of the top health foods: seafood. Healthy people who eat lots of seafood tend to have:

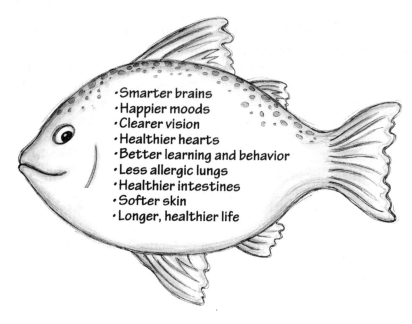

- Smarter brains
- Happier moods
- Clearer vision
- Healthier hearts
- Better learning and behavior
- Less allergic lungs
- Healthier intestines
- Softer skin
- Longer, healthier life

What Nutrients in Seafood Make It So Special?

In my search through hundreds of scientific articles, the health and longevity nutrient that kept getting scientists' top vote was the omega-3 fat found in fish oil. This discovery led to my book: *The Omega-3 Effect* (Little, Brown, 2012). A six-ounce fillet of wild Alaskan sockeye salmon, in my opinion, wins the top health food award because it packs the healthiest nutrients per calorie. While the omega-3s, vitamin D, protein, and calcium in seafood all deservedly get the most press, there is an overlooked, undervalued, and least understood nutrient in pink seafood that is one of nature's most powerful health foods –astaxanthin.

WHY SALMON ARE PINK

As I was fishing for knowledge on why seafood is one of the perfect health foods, I had the opportunity to go fishing with a real Alaskan fisherman, Randy Hartnell, owner of Vital Choice Seafood Company. While watching wild salmon swim mightily upstream during their marathon race, I was amazed at their strength and stamina to keep going without tiring. One morning while watching this salmon race, I asked Randy, "Why are salmon pink?" His answer led to the writing of this book.

Wild salmon feed on krill, small fish, and algae that are plentiful sources of the pink pigment astaxanthin. When that internal switch clicks on and drives these fish to leave the ocean and return to their birth river to spawn, as if guided by some internal GPS, they stop feeding during their vigorous journey upstream and live on their rich stores of fat and other nutrients accumulated for their race. Before their final sprint, the salmons' flesh turns a deeper pink because it stores astaxanthin from their diet. This powerful pink nutrient acts like an internal bodyguard to protect their body and keep it strong during the marathon. The harder they work, the pinker they get. As the fish digest their own stored fat for energy, a large dose of astaxanthin migrates into their flesh, turning it pinkish-red and giving the upstream racing fish their nickname, "red salmon."

Dr. Mother Nature Protects Against Muscle Fatigue

To give salmon the strength to swim upstream, their muscles undergo oxidative stress, which simply means lots of tissue wear-and-tear and muscle

Salmon swimming upstream.

fatigue. Think about it- these mighty fish continue swimming miles up raging rivers for up to seven days. What makes these heroic swimmers accomplish the greatest athletic feat in nature without damaging their muscles? The answer is: natural astaxanthin. Without this pink tissue protector, it's unlikely the salmon would reach their destination.

It's interesting that natural astaxanthin (an antioxidant) is found in the highest concentrations in the muscles that are stressed the hardest (that produce oxidants), such as the flesh of wild salmon. Isn't Dr. Mother Nature clever to instill the strongest antioxidant in the muscles that endure the highest oxidative stress?

Deep down I was concluding: If astaxanthin is fabulous for fish, could it also be fabulous for humans? The scientist in me wondered if the nutritional principle that applies to fish was what Mom used to preach about fruit and veggies: *the deeper the color, the better it is for you.* The reddish-pink color in wild salmon comes from the nutrient astaxanthin, one of nature's most powerful *antioxidants*. Antioxidants are nutrients that neutralize *oxidants*, wear-and-tear biochemicals the body produces not only during intense exercise, like the salmon run, but also as byproducts of the daily energy production during tissue growth and repair. This pink "medicine" is many times more potent than the antioxidant vitamins E and C. So, Mom's wisdom of putting more color on your plate pertained not only to fruits and vegetables, but also to fish.

Remember, **Astaxanthin = Antioxidant.**

FISHING IN HAWAII

Next, I went from pink fish to red ponds. In fishing through medical journals, I kept seeing the term "astaxanthin." So, I headed off to the Big Island of Hawaii to see the world's largest astaxanthin farm. My first impression as I gazed upon the twenty-five blood-red ponds was what a field of dreams this was. The eye-opening scene made me wonder: if fish eat this supernutrient from Dr. Mother Nature to stay healthy, and we eat the fish that eat the supernutrient, then it must be good for us. Three more visits to these red ponds confirmed what scientists say: Astaxanthin is one of the world's best-kept health secrets.

"The number one supplement that you've never heard of that you should be taking." – From the "Dr. Oz" show.

Aerial view of Astaxanthin farm, Kailua-Kona, Hawaii. Photo courtesy of Cyanotech Corporation.

Pinking Up Pale Salmon

Why are some salmon pinker than others? The nutritional wisdom "we are what we eat" also applies to salmon. One reason wild salmon are pinker than the farmed ones is that the wild ones eat natural astaxanthin; the farmed ones eat synthetic astaxanthin. Natural astaxanthin has deeper health effects; synthetic astaxanthin shows only *cosmetic* effects.

Suppose you were in the business of fish farming. Since the main goal of business is profit and not health, you would feed your fish the cheapest food that still makes them grow and look like real fish. Fish are picky eaters. They won't grow or even survive unless they eat at least some seafood. Their genes program them to be that way. These farmed fish make their way to the restaurant menu as "Atlantic salmon." Early on in your fish farming experiment you do a health check. You notice that some of your salmon are pale and not pink. These salmon won't sell because they don't look like real salmon. So you call your local fish doctor who sends you a "pill" to pink up your pale salmon to look healthier. That "pink up" pill is synthetic astaxanthin. The problem with this pseudo-pink stuff is that it is like other man-made, synthetic "food" – it may not be as healthful for the body as what Dr. Mother Nature made. A 4-ounce fillet of wild sockeye salmon, which is the best-known salmon in the Pacific, contains 4 milligrams of astaxanthin. You would have to eat two pounds of farmed Atlantic salmon to get that much astaxanthin.

Another concern is that synthetic astaxanthin is made from petrochemicals. That's right – pretty much the same stuff you put in the crankcase of your car. Not only is synthetic astaxanthin from a questionable source, but it's also a different animal in terms of how it works. For example, in one antioxidant experiment at Creighton University, natural astaxanthin from algae was twenty times more potent than synthetic astaxanthin (Bagchi, 2001).

See the Difference in Seafood

Next time you're in a supermarket that has a wide variety of seafood, compare a fillet of wild Pacific salmon to a farmed Atlantic salmon. The one that lived the wild life is pinker, almost red. The factory-made, pale pink color (synthetic astaxanthin) in the farmed fish doesn't behave in the body as well as the deeper-colored pink pigment (natural astaxanthin) in the fish the fisherman caught by hook or net.

A Tale of Two Fish: The Wild One and the Farmed One

Millions of years of recipe perfecting have produced the perfect food for growing perfect fish. The wild one, by some still mysterious primitive instinct, tells itself, "It's my time to find my birthplace, and I must swim there no matter what. Nothing will hold me back." But, like muscle training for marathons, the wild one knows she won't make it to her destination without storing up nutrients for the vigorous voyage. She must live off her own flesh that gives of itself for the good of getting there. So just before the race, the wild one goes on a feeding frenzy, gobbling up all the pink powerful foods (krill and astaxanthin-rich sea plants) she can to protect her muscles from self-cannibalizing during the upstream run. This is when fishermen try to catch the wild salmon: when they're at their pinkest, fattest, healthiest, and tastiest, just before they start to make their run upstream. The depth of their deep color – and their highest nutrient concentration – occurs while they keep themselves in a kind of natural holding tank days or weeks before their "go button" prompts them to swim for the fishy finish line.

The farmed one, on the other hand, sits around eating human-concocted recipes of food all day. Even if inclined to swim upstream and join the wild ones in the race, she can't escape from the pen to do so. So, the penned-up, less-perfect fish gets fatter from a substance that is less healthy for it than what the wild ones eat, and ultimately, the fish itself is less tasty when humans eat it.

Here's another mother-truism that you probably never understood even if you did listen: "You're not only what you eat, you're also what the animal eats." If you eat the wild one, you also eat those wild nutrients, and those nutrients become you. If you eat the farmed one, well, you get the picture. (See more about natural vs. synthetic astaxanthin in Chapter Five.) Wild fish have a healthier nutrient profile *because they eat real food*. We humans could learn that lesson from our sea friends.

From Colorful Fish to Colorful Humans

Now that you've followed these fish stories and value the pink pigment that keeps fish healthy, let's learn more about what astaxanthin is and what healthful effects it has on your body.

What works in nature can also works in humans. Could astaxanthin be as healthful for human bodies as it is for salmon? Yes! Since I'm an exercise addict, I realized that the more my body moves, the more astaxanthin I need. This was my logic:

- Dr. Mother Nature proved it's safe and effective for hard-working tissues
- My body is a hard-working tissue
- Science says it's healthful
- Common sense says it's healthful

Pass the astaxanthin, please!

ALL ABOUT ASTAXANTHIN: WHAT IT IS, WHAT IT DOES

What is astaxanthin and what makes it makes it so beneficial to the cells in our hard-working bodies? Astaxanthin is a *nutraceutical*, a nutrient that has proven pharmacological health benefits. Astaxanthin is correctly called "The King of Carotenoids." While you may not know what carotenoids are, you've probably eaten a few of them in the past twenty-four hours. Carotenoids are the pigments that give many of the healthy foods we eat their color: the lycopene in the red tomato, the zeaxanthin and lutein (you've probably heard about these for eye health) in the yellow corn, and the beta-carotene in the orange carrots. While there are over 700 different carotenoids in nature, you've probably only eaten a few of them.

Where in the world of nature is astaxanthin? Besides salmon gobbling it from the ocean's smorgasbord, it is also what gives pink flamingos their color. Flamingos eat algae that contain the carotenoid zeaxanthin and the orange carotenoid beta-carotene. Their bodies convert these carotenoids into the pinkish-red carotenoid astaxanthin. Among the plant life of the sea, algae is the most prevalent. The alga that produces astaxanthin is called *Haematococcus pluvialis*. Haematococcus algae are one-cell plants; they're like microscopic little red balls that can stay alive for years, surviving intense sunlight and other forces of nature such as the summer sun and winter cold. What makes them so strong is their accumulation of astaxanthin, also known as *"The Great Protector."* The powerful pigment turns the algae red and, in so doing, not only protects them from damage but enables a ready food supply for hungry fish.

Astaxanthin can be found in sea plants and sea animals. It is most prevalent in algae and phytoplankton, or sea plants. Any sea animal that has a reddish or pinkish color, such as salmon, trout, lobster, shrimp, and crab, contains natural astaxanthin. These seafood eat krill and other small sea organisms that eat astaxanthin-containing algae and plankton as a major part of their diets. The

7

sea animal that has the highest concentration of this king carotenoid is salmon. Astaxanthin concentrates in their muscles and makes them the endurance heroes of the animal world. If it weren't for astaxanthin, not only would the salmon cease to be a delectable delicacy on our tables, but they'd be pale, worn out, and tired all the time. Sounds like some humans who are antioxidant-deficient.

Now let's take a trip through the body, even as deep as the cellular level, to learn about what one of nature's most powerful and colorful pigments can do for you.

SCIENCE SAYS: ASTAXANTHIN IS AWESOME

While studies of the health effects of astaxanthin are still in their infancy, preliminary research shows that this pink powerhouse can:

- Keep the brain strong
- Keep eyes healthy
- Support the immune system
- Protect the skin from UV sun damage
- Keep the lining of blood vessels smooth
- Help blood lipids – HDL, triglycerides – stay in the normal range
- Support tissues undergoing wear and tear during intense exercise.

HOW WE STAY WELL: SHUN THE "STICKY STUFF"

When I was putting together my own health plan, I thoroughly researched the question: What exactly happens in the body that causes aches, stiffness, and poor health? Since I do my best thinking while swimming, I often say to myself: "I'll swim on that!" For me, the free-flowing movement of the body pushes aside distracting and competing thoughts and allows clearer thoughts to flow. It was on a swim that I came up with the term *"sticky stuff."*

I wanted a simple, memorable, teachable, yet accurate explanation of how our body parts break down and how we can keep them healthy. While health is dependent on many factors -- including genetics, family history, lifestyle, and diet -- the accumulation of "sticky stuff" in tissues plays a key role in a variety of health conditions and accelerated aging.

As a physician I try to explain health and wellness in simple terms. My patients like my "sticky stuff" explanation: When you accumulate too much sticky stuff in your body, you're likely to feel pain, stiffness, and swelling, and eventually seek the help of doctors to find relief. When you keep excess sticky stuff out of your body and adopt a sensible wellness plan, you increase your chances of staying healthy – at all ages.

The good news is that consuming astaxanthin, the nutrient naturally found in seafood, is one weapon against accumulating too much sticky stuff in our bodies.

Sticky Stuff Story

What actually is this sticky stuff? Medically speaking, this sticky stuff begins with the process of *oxidation* – the "shun" word. Simply speaking, for good health and longevity, shun the "shun." Our bodies are oxygen-burning machines. Every minute, countless biochemical reactions throughout the body generate trillions of particles of "exhaust." These sticky particles of exhaust are known as *oxidants*, also called free radicals. Trillions of times a day these oxidants hit our tissues like a steady rainfall rusting away our cells. The rusting caused by these oxidants leads to wear and tear associated with aging, such as stiff joints, sore muscles, blurry vision, and wrinkled skin.

Antioxidants Are Anti-Sticky Stuff

Normally, our bodies soak up these oxidants by producing anti-rust chemicals called *antioxidants*. But when the body builds up more oxidants than antioxidants, rust accumulates and increases the wear and tear on our tissues. Unfortunately, as we age, our bodies tend to produce fewer antioxidants. So, as we get older, we need to eat more foods that are rich in antioxidants.

ASTAXANTHIN SUPPORTS HEALTHY AGING

Once upon a time it was thought that the ailments of aging were mostly due to tissue wear and tear over time. The aging body was viewed as a machine that simply wears out because of all the parts rubbing together: blood flowing through blood vessels eventually wears out the vessel lining; joints being jarred all day wear out the joint lining; and skin exposed to too much sunshine gets wrinkled. This simplistic explanation of aging is only partially true.

An updated explanation of aging is that the tissue wear and tear causes the "shun" word – *oxidation*, which leads to sticky stuff, which triggers even more tissue damage, which causes tissue to get too old too fast, which is what happens when we age too quickly. A downward spiral. (Another "shun," *glycation*, happens when sticky stuff accumulates from eating too much added sugar.) Astaxanthin, an antioxidant, supports healthy aging by helping to decrease the "shuns."

> ### — The AAA Effect of Aging —
> When we *Age*, our bodies produce less *Antioxidants*, so we need more *Astaxanthin*.

The astaxanthin effect is particularly healthful for seniors (see Chapter Three on the Head-to-Toe Effects of Astaxanthin), especially for the skin:

- The anti-wrinkle effect
- The anti–sun damage effect
- The antioxidant effect

Sticky Stuff Cycle:

Sticky stuff accumulates in tissues – oxidizing the tissues then wearing them out.

Tissues are worn out – Because the tissues are now damaged by the sticky stuff, the immune system doesn't recognize them. Our protector cells and germ-fighting cells wrongly conclude that because this sticky stuff doesn't belong here, we must attack it.

Immune response – Overactive immune response causes more sticky stuff to build up in the tissues, which leads to them wearing out. Remember, sticky stuff is another term for oxidants. To fight oxidants, you need antioxidants. But other than from our own bodies, where do oxidants come from?

Where We Get Oxidants??

The five main sources of excess oxidants are:

1. Normal body chemical reactions, or metabolism
2. Excessive ultraviolet-ray sun exposure
3. Air pollutants
4. Overexertion, which generates too much muscle "exhaust" for our bodies' antioxidants to handle
5. Overworked immune system, which attacks oxidation but in doing so produces even more oxidants.

Revitalizing Aging Tissues

When your body is in balance, you tend to enjoy health and longevity. When your body is out of balance, you are more prone to getting sick. So, the secret to good health is keeping your body in balance, which means shun the "shun" of excess oxidation.

To understand what happens when the body breaks down and you start feeling achy, stiff, and tired, let's use the road repair analogy. Suppose there was a road well-traveled in your body – say, the lining of a blood vessel that has blood (traffic) flowing through it, or even the lining of a joint, such as a knee joint, that's flexed thousands of times a day. There is a lot of rubbing that goes on in these tissues, so that eventually the lining gets rough. Your body has a marvelous built-in maintenance system. It sends out the message that the road is rough and needs repair. Maintenance engineers are deployed. These are actually repair cells that go by various biochemical names, such as cytokines. Their job is to mobilize the repair crew to fix the wear and tear on the road – fill the potholes and smooth the surface. Yet, sometimes these overzealous maintenance engineers overreact, or over-repair. The potholes are overfilled, leaving a bumpy road and other rough surfaces.

So, the body mobilizes even more maintenance engineers, but during their repair they add more buildup of sticky stuff, or bumps, in the road. Eventually, the road gets so bumpy that traffic comes to a halt (e.g., a clot forms in the artery). Over time, this leads to dysfunction of the organs whose blood is supplied by the artery (resulting in a stroke or heart attack); or joints get painfully frozen and need repair, such as knee or hip replacement. This simple but underappreciated cycle of breakdown and repair highlights the importance of maintaining healthy balance in our diet and lifestyles. Balance is one of the top secrets to good health

Get the Chemical Picture

What makes the astaxanthin molecule so special? The best antioxidants are those that get into the most tissues. Some antioxidants get into fat tissue, some get into blood, and some get into muscle. Here's where astaxanthin shines. It gets into all three. Astaxanthin enjoys a chemical perk that sets it above the other carotenoids. It has an extra biochemical formula called a hydroxyl group attached to both ends of the molecule. This enables astaxanthin to work its way into tissues better than the other antioxidants.

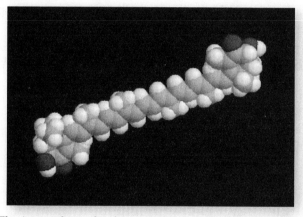

The Astaxanthin molecule's special hydroxyl group at both ends.

2
MEET ASTAXANTHIN THE ANTIOXIDANT

I had often heard the terms "oxidation" and "antioxidants," yet like so much biochemical stuff you learn, you forget it until it has personal meaning. Science says: "People who eat the most antioxidants enjoy the best health." When I did a scientific literature search for "antioxidants," astaxanthin kept popping up, and it was easy to remember:

The Astaxanthin Effect = The Antioxidant Effect

So, to fully understand the astaxanthin effect, let's first understand the antioxidant effect.

Astaxanthin is oxidant quenching.

ENJOY THE ANTIOXIDANT EFFECT

Too Many Oxidants

Every cell in our body, trillions of them, are microscopic engines burning the fuel (food) we eat to generate energy to grow, repair, move, think, and so on. Just as a car engine burning fuel generates exhaust, our body generates exhaust, called *oxidants*, from the fuel – oxygen that the body burns. Accumulation of

14

oxidants can lead to early aging and many health conditions. Oxidation, a natural biochemical process, is our body's way of converting food and oxygen into energy. We can't live without it. But too much oxidation can hurt us. Excess oxidation throws the body out of biochemical balance.

Time to "Anti" These Oxidants

Now that you've learned a bit about oxidants, let's go deeper into your metabolism to understand how excess oxidants harm your health and why you need to "anti" these oxidants. Oxidation, like the burning of fuel (such as in a car engine), is like a mini explosion that occurs trillions of times every minute throughout your tissues. When these energy-producing explosions occur, they give off biochemical stuff called *oxidants*. They are also known as *free radicals,* so-called because they behave in the body like a bunch of out-of- control cell damagers. Basically, they are free electrons. In simple yet scientific language, the oxidant effect on our tissues is known as "hits." The hardest-working tissues, such as the brain and eyes, take the most hits. According to University of California antioxidant researcher Dr. Lester Packer, author of *The Antioxidant Miracle*, each cell in our body takes around 10,000 oxidative hits daily. Multiply that by trillions of cells and each day your body takes a lot of "hits."

To help tissues heal, these hits need their buddies – anti-hits, or antioxidants – to keep tissues in antioxidant balance. Otherwise, these oxidants, like mini jackhammers, will keep hitting away at tissues, causing damage. Once the oxidants go haywire and oxidation begins, a chain reaction can occur that generates more oxidants, causing a downward spiral of one health problem leading to another.

An unfair quirk of aging is that as we get older, most of us continue to produce just as many oxidants, but our bodies decrease their production of antioxidants. When oxidants equal antioxidants, you tend to remain healthy. When oxidants outweigh antioxidants, you are out of balance, and wear and tear can accumulate, leading to accelerated aging.

Like a well-designed car engine, the body can handle a normal amount of exhaust, or oxidants. A beautiful balance of wear, tear, and repair goes on 24/7 in every part of your body. Yet, our modern lifestyle and diet have caused many of us to have a body out of balance: wear and tear occurs faster than repair. The pollution that goes into our bodies, especially through the chemical air we breathe and the chemical food we eat, shifts the balance toward excess oxidation, or more wear and tear than repair. In a nutshell, we wear out too soon.

Shift your balance toward eating more antioxidants
and decreasing your exposure to oxidants.

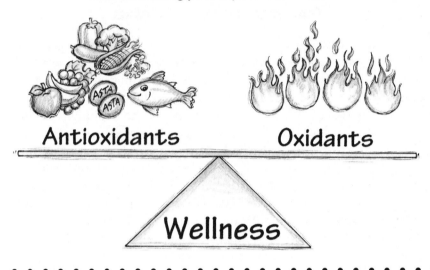

Biochemically, sticky stuff, or oxidants, is the waste product of cellular metabolism. The anti–sticky stuff is known as antioxidants. These antioxidants neutralize, or suck up, the oxidants before the oxidants can do their tissue damage. Think of oxidants as trillions of tiny hot potatoes that need to be passed from one antioxidant to another until they cool off – before they "burn" the tissues.

Aging Is "Rusting"

Back to the car analogy. What happens if you expose metal to the elements over a long time? It oxidizes, or rusts. So, if you put paint (an antioxidant) on the metal, it keeps the metal "young." While it doesn't sound so sexy, aging is rusting.

Any health plan needs to begin with what you can do and eat to neutralize the oxidants that are formed every millisecond in your body. Simply put, accelerated aging is caused by oxidants; therefore, eat more antioxidants. You can do that!

The older you get, the more antioxidants you need to eat.

The Avocado Analogy

Here's a graphic home exercise to appreciate how antioxidants, such as astaxanthin, help neutralize damage from free radicals. Take either one of the "A" plants that are health foods: apples or avocados. I prefer to do this demonstration with an avocado to make my point, since another nutritional principle is that the more healthy fats a food contains, the quicker it spoils (oxidizes). Halve the avocado. Pour lemon juice (contains antioxidants, mainly vitamin C and other biological protectors called bioflavonoids) on one half and leave the other half exposed to the air unprotected. Reexamine your experiment in around six hours. The antioxidant-protected avocado half looks fresh, young, and healthy. The unprotected half looks unhealthy, wrinkled, old, and even rusty. Without antioxidant protection, the avocado deteriorated.

The avocado analogy is what happens to our bodies when we either expose our tissues to oxidation or don't eat enough antioxidants. Just as the avocado rusts from oxidants, so can oxidants "rust" our tissues, skin, muscles, joints, eyes, and arteries. (You will learn more about this later.) The illustrations below demonstrate the shift in your wellness that can occur as you increase or decrease the intake of antioxidants.

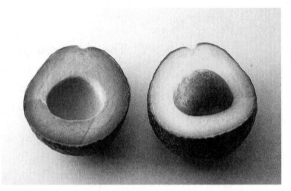

Avocado halves – unprotected (left) and antioxidant-protected (right).

"Shuns" Don't Show Up Overnight

The "shuns" build up over a long time. Notice the artery below. I use an artery as one of the main tissues affected by the "shuns." In fact, I directed my health plan toward healthy arteries, intuitively knowing that every organ of the body is only as healthy as the blood vessels supplying it. Therefore, if I have healthy blood vessels, I'll have healthy tissues and healthy organs.

WHY ASTAXANTHIN IS THE MOST POWERFUL ANTIOXIDANT

Now that you understand what antioxidants are and what they do, let's choose the best ones for you.

An Imaginary Antioxidant Contest

Suppose all the top antioxidants got together for a contest. Let's call it a health-lifting competition. Many of the contestants were familiar favorites:

- A, B, C, D, E team (the vitamins)
- Flavonoids (represented by the berries)
- Carotenoids (represented by the yellow, orange, and red vegetables)
- A relative newcomer to the health food scene (astaxanthin).

"Who is the strongest antioxidant?" each contestant was asked. All claimed to be the leader, yet the judges shouted, "Show me the science." When the judges examined the scientific literature, they came to two conclusions: (1) "Astaxanthin beats all of the other teams and (2) "You're all winners!" If Dr. Mother Nature were a real scientist she would conclude: "You all are winners because nowhere in nature do nutrients act alone. They always partner with one another for a synergistic effect." (See Chapter Five for an explanation of the nutritional principle of synergy.)

How do scientists measure antioxidants and their effect on the body? The current blood tests are not precise. At best, scientists observe the effects of antioxidants in laboratory test tubes or in experimental animals, such as rats and mice. Then they make the leap of faith and apply lab and animal results to the human body.

Tests Show Astaxanthin Overpowers Many Antioxidants

This does not mean you should take astaxanthin instead of other antioxidants, especially those found in fruits and vegetables, since all have their unique effects. Notice from the graph below that natural astaxanthin is over 5,000 times stronger than vitamin C and much stronger than two other well-known antioxidants, co-enzyme Q10 and green tea catechins.

If you're interested in how these antioxidant effects are measured, here's one way it's done. You measure the antioxidant's singlet oxygen neutralizing capability, meaning how powerfully it neutralizes one of the most damaging oxidants, called singlet oxygen. In biochemical speak, this neutralization is called "quenching," sort of like putting the fire out.

The Astaxanthin Effect in the Body

While it's important to measure an antioxidant's strength in a laboratory, what's even more important is how it behaves in the body. Four important points to consider are whether the antioxidant can:

- Cross the blood-retinal barrier to bring antioxidant protection to the eyes. Astaxanthin – *yes*.
- Protect both fat-soluble and water-soluble parts of the cell, the parts

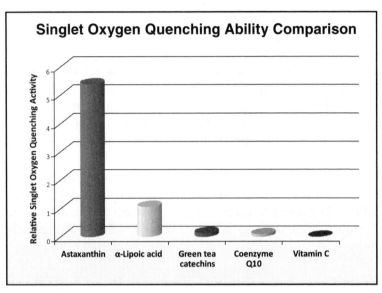

SOURCE: *Nishida et al., 2007.*
NOTE: *In vitro laboratory testing was used to compare the antioxidants.*

most vulnerable to damage. Astaxanthin – *yes*.
- Get into and protect muscle tissue. Astaxanthin – *yes*.
- Become a pro-oxidant instead of an antioxidant and actually cause oxidation and subsequent cell damage. Astaxanthin– *no*!

An example of an antioxidant becoming a pro-oxidant is ingestion of excess iron.

You'll notice that the graphs to the right show different values according to different laboratory tests that are used. The three most popular tests are the singlet oxygen quenching test, the oxygen free radical scavenging test, and the ORAC test (oxygen radical absorbing capacity). For example, in the test measuring singlet oxygen quenching, astaxanthin proved to be 550 times stronger than vitamin E. In the test measuring free radical scavenging, astaxanthin was twelve times stronger than vitamin E. This is why it can be misleading to rely on a single test to measure an antioxidant's strength and why the popular ORAC test (not shown) is not the only one to consider. It's interesting that in all three antioxidant tests, astaxanthin scored high above all the rest. In the ORAC test, astaxanthin proved to be much more powerful than other common antioxidants.

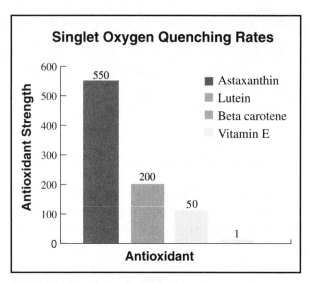

SOURCE: Shimidzu et al., 1996.

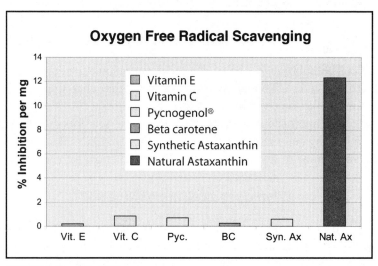

SOURCE: Bagchi, 2001.

LEARN THE CELLULAR SECRET TO HEALTH

A medical truism is that the body is only as healthy as each cell in it. The healthier the cell, the healthier the whole body – that is, the whole is only as healthy as the sum of its parts. Since there are around thirteen trillion cells in the body, that's a lot of parts to keep healthy. Here's how astaxanthin helps keep cells healthy.

Hardest-Working Parts of the Cell

Remember the principle of tissue wear and tear: the hardest-working tissues produce the most oxidants and need the most antioxidant protection. The following areas of the cell are hard-working and need the most antioxidants:

- The cell membrane, which is the outer lipid or fat layer of the cell
- The DNA, or genetic code within the cell, that dictates how correctly the cell replicates itself when it needs to multiply, grow, or replace itself when it wears out
- The mitochondria, which are like tiny batteries within the cell producing the energy the cell needs to do its work.

Protect Your Cell Membranes

The cell membrane is one of the most marvelous structures in the entire body. Picture it like a flexible covering, or skin, that encompasses all the cell contents, sort of like the covering of a ball.

One of the microscopic secrets of wellness that few folks appreciate is how important a healthy cell membrane is to overall health. Let's take a trip across this marvelous membrane to understand how it works and how we can keep it healthy.

What does a cell membrane do?

Each cell in your body is like your body's energy and nutrient bank. There is a lot going on inside there that needs protection by the membrane. The cell membrane has two major functions: (1) It transports nutrients from the outside to the inside of the cell, that is, from the bloodstream into the cell. (2) It protects the vital cellular contents from leaking out.

So a microscopic secret to health is to have healthy cell membranes that effectively transport and protect. Healthy cell membranes are the basis of internal health. Taking care of your cell health is where internal health begins.

Are your cell membranes healthy?

Dr. Bill

What is a cell membrane made of?

The cell membrane is a marvelous structure. It's composed of roughly half protein and half fat. One of the most metabolically magnificent construction projects within the body is how the fat molecules within the cell membrane line up to both protect and transport. Remember, oil and water don't mix, which is why the membrane contains a lot of fat, allowing the cell to sit next to a stream of liquid (blood vessels and tissue fluids) without dissolving. Drop an all-sugar piece of candy into a glass of water and see how quickly it dissolves. Now take a chocolate chip and see how it dissolves more slowly (it contains some fat). Now take that chocolate chip and remove all the sugar, add protein and certain fats. It wouldn't dissolve at all. It's a good thing our cell membranes are designed as they are, otherwise our bodies would quickly "dissolve."

23

Here's how astaxanthin protects the cell membrane

Fat is the tissue most vulnerable to oxidation, that "shun" producing process. Here's how to protect it. I remember that smelly fillet of fatty fish that turned rancid (oxidized) after I mistakenly left it on top of our freezer overnight. To protect this precious membrane fat, you need the precious pink nutrient astaxanthin. If astaxanthin could talk, it would say, "I protect your fat." Here's why membrane fat and astaxanthin are best buddies.

Astaxanthin protects cell membranes better than many other antioxidants, including the popular vitamin C and beta carotene. Fat-molecule membranes, called cellular membrane lipids, line up vertically like a picket fence, enclosing the precious contents of the hard-working, growing, and reproducing animals inside the cell contents. The membrane, or fence, lets in what the animals need, keeps out what they don't need, and transports back out, shall we say, the "stuff" that builds up within the cellular pasture.

What makes astaxanthin such a special cell-membrane protector? It enjoys a double biochemical property: it is both lipophilic, meaning it loves fatty tissue, and hydrophilic, meaning it loves water. This enables it to work in tissue that contains both fat and water – the structure of the cell. *Both* surfaces of a cell's fatty membrane need protection: the surface that is in contact with the fluids outside the cell, which keeps it from dissolving into the bloodstream, and the surface that's in contact with the fluid inside the cell. The antioxidant vitamin C helps protect the outside fat molecules because the outside surface is water-soluble, but it can't protect the inside, or fat-soluble layer of the cell. So, it only does half the job. Beta-carotene does just the reverse. Because it can penetrate fat, beta-carotene protects the fat inside the cell membrane from oxidation, but it doesn't protect the water-soluble surface.

Astaxanthin does both. Because of its unique biochemical structure, it spans the entire membrane to protect both layers – lipophilic and hydrophilic. Most antioxidants are soluble in either fat or water, but not both. Astaxanthin is a unique antioxidant in that it is both fat and water soluble, able to protect both the outside and inside of the cell.

As you will learn throughout this book, some of the most vital tissues in your body are fat: the fatty layer in cell membranes, the fatty layer that wraps around nerve cells, the retinal tissue of the eye, and, of course, the skin. Like a security guard, astaxanthin easily penetrates fatty tissue and helps protect against oxidants that could get into the cell membrane and damage it.

If the cell membrane could talk, it would say to astaxanthin: "We're made for each other!"

Astaxanthin protects the inside of the cell

Astaxanthin is aptly called the outside/inside cell health protector. The micro factory of a cell works 24/7, producing oxidants (waste) that are neutralized by the cell's own antioxidants. One of the most well-known oxidants the cell produces is superoxide, which is neutralized by a natural cellular enzyme called *superoxide dismutase*, better known as SOD. But, as cell biologists say, "accidents happen" within the hard-working cellular factory. Some of these oxidants escape into the cellular machinery before the SOD police can grab them, bind them, and prevent them from doing harm, such as oxidizing, or rusting, the cells. But, as astaxanthin is present inside and outside the cell, it can grab and neutralize excess oxidants and, in particular, has been shown to be a very effective and neutralizing superoxide..

Another protective property of astaxanthin is its *double bonds*, the biochemical structure that makes a molecule a powerful antioxidant. Think 007, Bond – Double Bonds. Double bonds mean they have an extra parking space for any free wayward electrons (oxidants) to park. They act like magnets to grab onto the oxidants and get them out of circulation.

The mighty mitochondria

The new science on aging tells us that the micro factories within each cell – called mitochondria – rust and wear out too fast. Mitochondria are some of the hardest-working molecules in the body. Remember, the hardest-working molecules generate the most oxidants and therefore need the most antioxidant protection.

Preserving the life of the cellular machinery is the main key to longevity. Healthy aging, therefore, is keeping the cellular machinery protected from rust. This is what powerful antioxidants, like astaxanthin, do to combat cellular aging. Take-home message: The healthier each cell, the healthier your whole body.

3

THE HEAD-TO-TOE EFFECTS OF ASTAXANTHIN

Let's take a trip through your body to see how astaxanthin supports your most vital organs. We'll start with how astaxanthin supports the blood vessels that lead to the heart, then we'll look at astaxanthin's nourishment to the brain, and then we'll see how it protects our eyes. Next we talk about astaxanthin's restorative effects on overused joints, muscles, and tendons in workers and athletes, and we'll describe how it supports the immune system. We'll conclude with the glowing skin effects of astaxanthin.

BE HEART SMART

Deep within your cardiovascular system lies one of the top secrets to health. Its discovery won the Nobel Prize.

A Night to Remember

As I was searching for the secrets to health, I was fortunate to learn from wise scientist friends and the health discoveries they made. As a physician, I believe that our bodies are designed to make most of the internal medicines we need. But where in the body is this internal pharmacy and what medicines does it make? The answer to this secret to health came one night at our home when Nobel laureate Dr. Lou Ignarro was our guest for dinner. Dr. Lou taught me where this pharmacy was and what it did.

Save Your Silver Lining

You're about to learn what "silver lining" means to your health, where it is in your body, and how you can keep it healthy. A medical truism is: "You're only as healthy as your blood vessels," because the health of every organ depends on the nourishment it gets through its blood vessels. The lining of your blood vessels, known medically as the endothelium, is what I call the "silver lining" because it is the key to health and longevity.

Your endothelium is a one-cell-layer tissue that lines the inside of your blood vessels. Think of it like wallpaper on the walls of your vessels. Yet, unlike wallpaper, it doesn't just sit there and look pretty. It does something. In fact, your endothelium is the largest hormone-producing tissue in your body. If you were to open up all your blood vessels and lay them out flat, your endothelium would occupy a surface area larger than several tennis courts.

The endothelium is one of your most important systems, yet many people don't fully appreciate what it is and what it does. Picture your endothelial lining containing millions of microscopic "medicine bottles" (biochemically called secretory cells). A healthy person who eats lots of antioxidants like fruits, vegetables, and seafood enjoys less buildup of sticky stuff (oxidants) on the endothelium. So the endothelial "pharmacy" stays open: *wellness*.

Endothelial Pharmacy OPEN

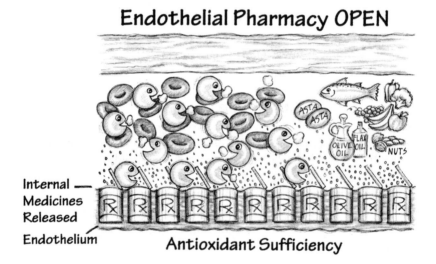

Internal Medicines Released

Endothelium

Antioxidant Sufficiency

During our "scientist and simpleton" dialogue, as Lou was explaining his research in heavy biochemical terms, I was drawing my medicine bottle explanation. That's how I came up with the teaching tool of the endothelium being your personal pharmacy within, and how the buildup of sticky stuff keeps the pharmacy closed.

Endothelial Pharmacy CLOSED

Sticky
Stuff

Internal
Medicines
Blocked

Endothelium

Antioxidant Insufficiency

Care for Your Personal Pharmacy

Consider your endothelium as your own internal drugstore churning out natural "internal medicines" 24/7. Yet, unlike the pharmaceuticals you buy, these internal medicines are custom-made just for you, released into your bloodstream at the right time, in the right dose, and without any side effects, and they're free.

The cells of the endothelium are like closely placed tiles on a countertop. But again, these tiles don't just sit there, they do something. Each tile is its own endocrine organ, like millions of microscopic medicine bottles. The endothelium is like the chemical command center of the blood vessels, telling them when to open up and deliver more blood to hard-working tissues, such as when the intestines need more blood flow after eating and when the muscles need more blood flow during exercise. In some ways, the endothelium is like having your very own internal medicine doctor inside your body, constantly sensing your medical needs and responding by dispensing the right medicine. In chemical speak, these small molecules and proteins within the endothelial cells ("medicine bottles") are known as biologically active substances.

What the "Medicines" in the Silver Lining Are

There are around 23 known chemicals, or "medicines," within those tiny "medicine bottles" naturally contained in your endothelium. There are "medicines" to regulate blood pressure and blood clotting, "medicines" to elevate the "lows" and mellow your moods, and "medicines" to heal your hurts.

Think of these medicine bottles inside the endothelium as millions of tiny pharmacists. The better you feed and care for these pharmacists within, the better they dispense the medicines you need.

Healthy Endothelial Function Is the Key to Good Health

Endothelial health contributes to cardiovascular health. Here's the simplest explanation for cardiovascular health you've ever heard:

Keep the sticky stuff off your silver lining.

Keep sticky fats and other sticky stuff off your vascular highways. Eat a diet that is high in vegetables, fruits, and seafood, and low in refined sugars.

If I were asked to describe the one goal or change a person could make to enjoy healthy organs and longevity it would be: Keep your endothelium healthy. Or even more basic: Keep the sticky stuff off your endothelium. As I was putting together my own health plan, I realized that one of my top goals for health and longevity was to keep my endothelial pharmacy open and releasing medicines as long as I can. Since every organ in the body is only as healthy as the blood vessels nourishing it, total body health begins with a healthy endothelium.

Four Ways Astaxanthin Can Help Your Heart

Astaxanthin is a powerful antioxidant. Here are four reasons why astaxanthin can keep your heart strong:

• ***Astaxanthin keeps the endothelium smooth.*** Having a healthy silver lining is like painting a highway with a non-stick surface so that the cars (blood cells) can keep moving quickly and not cause a traffic jam. In an animal study, Japanese researchers discovered that administration of astaxanthin for fourteen days resulted in improved blood pressure. The general conclusion was that astaxanthin could help relax blood vessels, resulting in lower blood pressure (Hussein et al., 2005, 2006). Although the results cannot be extrapolated to humans, they provide an important foundation for future clinical research in humans.

• ***Astaxanthin supports healthy blood flow.*** Research in experimental animals shows that astaxanthin can help the heart maintain normal blood flow (Gross, 2005; Nakao, 2010). A study on adults with metabolic syndrome showed that astaxanthin supported healthy blood by promoting arterial health (Satoh et al., 2009; Fassett et al., 2008). In a study of human volunteers, participants supplemented with 6 milligrams of astaxanthin per day for only ten days showed a significant improvement in blood flow (Miyawaki et al., 2008).

• ***Astaxanthin keeps CRP levels at a normal range.*** One of the most common blood markers scientists use to detect stress in the body is called CRP, or C-reactive protein. CRP is produced in the liver and released into the bloodstream when the body is fighting the aches and pains associated excess oxidation. A 2006 human clinical study conducted by The Health Research and Studies Center in Los Altos,

California, studied 25 people for eight weeks. Sixteen people were given natural Hawaiian astaxanthin and nine received a placebo. The group given astaxanthin experienced a 20 percent reduction in CRP levels in just eight weeks, whereas the placebo group had an increase in CRP levels (Spiller et al., 2006).

•   ***Astaxanthin helps keep blood lipid levels at a normal range.*** Studies show that natural astaxanthin can support normal LDL and triglyceride levels, and HDL (Awamoto et al., 2000). A 2010 study from Japan on humans showed that adults who took astaxanthin supplements had an improved lipid profile, namely decreased triglycerides (sticky fats) and increased HDL cholesterol (non-sticky fats). Blood levels of the hormone adiponectin were also higher in the people who took astaxanthin. Adiponectin is a newly discovered natural hormone that helps promote healthy blood sugar and blood fat levels. Optimal results were found at 12 milligrams of astaxanthin per day (Yoshida et al., 2010). An earlier study of humans taking astaxanthin supplements had also demonstrated that astaxanthin supports healthy cholesterol levels and triglyceride levels already within the normal range (Trimeks, 2003).

Conclusion: Astaxanthin Protects the Heart

A study by Harvard researchers revealed another clue to how astaxanthin helps hearts. Remember the anti-inflammatory drug Vioxx® that was pulled because it increased the risk of heart disease? The researchers discovered that Vioxx harmed hearts by increasing the susceptibility of heart cell lipid membranes and LDL cholesterol to oxidation. In other words, it behaved like a pro-oxidant. But the researchers went on to find that astaxanthin, a powerful antioxidant, blocked the oxidation effect of Vioxx during experiments. Simply put, Vioxx attacked tissue fats; astaxanthin provided antioxidant protection (Mason et al., 2006).

In fact, cardiologists now conclude that most of the cardiovascular benefits of astaxanthin are due to its antioxidant effects. In a nutshell, it decreases oxidative stress on the heart and blood vessels to support cardiovascular health (Pashkow et al., 2008).

Good science and good sense usually go together. Astaxanthin is a powerful antioxidant. Go red!

HOW ASTAXANTHIN KEEPS THE BRAIN HEALTHY

Remember this health insight: Because the hardest-working tissues produce the most oxidants, they need the most antioxidants. Three unique features of the brain make it a high-need tissue for antioxidants like astaxanthin.

• **The brain is the hardest-working tissue of the body.** It uses 25 percent of the carbohydrate energy you eat and burns 20 percent of the oxygen you breathe. Yet, your brain makes up only about 2 percent of your body weight.

• **The brain is 60 percent fat.** Fatty tissue is the most vulnerable to oxidation or the excessive wear and tear known as "oxidative stress," even more so than high-protein tissue like muscle. You might say that the brain, especially as we age, gets stressed out.

• **The blood-brain barrier (BBB) often weakens with age.** Although the BBB screens out chemicals, called neurotoxins, that may harm brain tissue, the BBB wears down with age, and neurotoxins get through. The aging brain is also less able to mute the effects of high-stress hormones, a condition called gluconeurotoxicity. And, the aging brain is less able to repair itself and grow new brain tissue.

Because your brain is such an important – and vulnerable – organ, a new term has been coined for nutrients that protect nerve tissue, *neuroprotectants*.

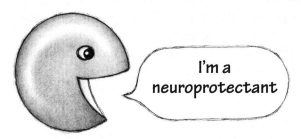

I'm a neuroprotectant

For obvious reasons, before supplements or drugs are used in humans, they are tested on experimental animals to see if they are safe and effective. The astaxanthin effect on nerve tissue protection has passed the animal testing phase, and the next phase is human testing. Here's a summary of what science says about astaxanthin as a neuroprotectant.

33

Supports Cognitive Health During Aging

Ten seniors with age-related forgetfulness received 12 milligrams of Natural Astaxanthin daily for twelve weeks. Astaxanthin was shown to be effective in promoting cognitive function and psychomotor function (Satoh et al., 2009).

Lessens Oxidative Stress to the Brain

After twelve weeks of taking 6 or 12 milligrams daily of Natural Astaxanthin, human volunteers were found to have a decreased blood level of oxidants called *phospholipid hydroperoxides* (sticky stuff). A laboratory study from the Departments of Surgery and Pharmacology at the University of Pittsburgh School of Medicine set up a mouse model and found that astaxanthin offers neuroprotection against oxidative stress (Lee et al., 2011). Researchers at the Nagoya University in Japan showed that human brain cells (in a laboratory dish) damaged by oxidation showed less damage when they had been pretreated with astaxanthin, leading the researchers to conclude that astaxanthin is a "natural brain food" for oxidative stress (Liu et al., 2009; see also Wang et al., 2010; Chan et al., 2009; and Kim et al., 2009).

Protects Healthy Nerve Tissue

The military researches high doses of antioxidants to see how antioxidants can support brain health following combat (Deuster et al., 2009). Again, science and common sense go together. If brain tissue under stress builds up oxidants, the antidote should be a powerful antioxidant.

Damaged nerve tissue, such as that incurred by a traumatic brain injury or a stroke, can repair itself by producing a type of neurological stem cell called a *neuroprogenitor cell* (NPC). A fascinating university study from Korea showed that proliferation of NPCs in experimentally damaged nerve tissue in mice increased when the healthy tissue had been supported by the antioxidant effects of astaxanthin. Researchers think that when a cell is damaged it sends out signals to other cells that it's in need of repair. The NPCs are like front-line medics summoned to the battle scene to treat traumatic brain injuries before they become fatal (Kim, 2010).

Researchers at the National Institutes of Health came to similar conclusions. Working with rats, they showed that brain tissue supported by the antioxidant

34

effects of astaxanthin led to less ischemic (stroke) damage than healthy brain tissue not supported by astaxanthin (Shen et al., 2009; see also Hussein et al., 2005; Kudo et al., 2002).

ASTAXANTHIN KEEPS YOUR EYES STRONG

Astaxanthin Crosses the Blood-Retinal Barrier to Protect the Retina

A fascinating tribute to the architecture of the human body is eye health, which can be drastically affected by potential toxins in the bloodstream. The eyes are protected by a one-cell-thick wrapping called the blood-retinal barrier (BRB). In studying astaxanthin's effect on the eyes, researchers wondered if astaxanthin had the ability to cross the BRB, since many antioxidants cannot. Way back in the 1940's, Drs. Rene Grangaud and Renee Massonet in their doctoral research at the University of Lyon in France demonstrated that astaxanthin does indeed cross the BRB and into eye tissue. By feeding laboratory rats astaxanthin, they showed that once astaxanthin reaches the eyes, it acts as a strong retinal protectant (Grangaud, 1951 and Massonet, 1958).

Astaxanthin Is "See" Food

The retina of your eyes, the back part of your eyes that functions as a projection screen, is an extension of brain tissue. Therefore, what's good for the brain should be good for the eyes. Science agrees. And, like brain tissue, retinal tissue is one of the most fatty and hardest-working tissues in the body. Both brain cells and eye cells are what I call high-need cells, those hard-working (high-metabolic) cells that need more antioxidant protection. The retina and especially the sensitive macula on the retina function like a camera that captures the photo of, say, your beautiful baby or grandbaby. The tissue is composed mainly of fat and a network of blood vessels highly affected by oxidation. In fact, retinal tissue has some of the highest metabolic rates of any tissue in the body. Because of its high rate of energy use, it generates lots of waste products, or oxidants, so it needs a powerful antioxidant. Let's focus on how astaxanthin can help protect your eye health.

Imagine your retina taking millions of snapshots in a millisecond, then processing these into a composite photo of your smiling baby. Fortunately, the eyeball is one of the easiest tissues to examine. Using an ophthalmoscope to look through that little black opening in your iris, your doctor can examine the back of your eyeball – the retina – and with frequent observations see how it responds to your healthy dietary changes, such as adding antioxidants. This is why many people, like me, notice clearer vision after adding antioxidants to their diet.

A well-known xanthin – zeaxanthin – has long been recognized as an important nutrient for eye health. Studies have shown that zeaxanthin, found mainly in dark-green leafy vegetables, corn, peppers, and egg yolk, is an essential nutrient that gets into the eye tissues, especially the lens and retina, to protect them against the aging effects of excessive ultraviolet light radiation.

Zeaxanthin and its buddy lutein, both carotenoids, protect macular pigments in the retina necessary for retinal health and clear vision. Zeaxanthin absorbs harmful blue light and reduces retinal oxidation, which is one reason that people who ingest more lutein and zeaxanthin have a lower incidence of age-related macular degeneration, the most common cause of vision loss. Until recently, lutein and zeaxanthin were thought to be the only carotenoids in our diet that could protect the lens and retina of the eye. Astaxanthin is the newcomer to the xanthin eye-protection family.

I help you see better.

Science Says More About Astaxanthin for Eye Health

Scientists conclude that astaxanthin acts as a "see food" by its powerful antioxidant properties, working in tissues most vulnerable to free-radical oxidation. As an additional perk, astaxanthin can increase blood flow to the retina, further protecting the eyes.

Protects tired eyes

Eyestrain or eye fatigue is a malady of modern computer life. Just as other tissues of the body get overworked, the eyes get tired. This is why antioxidant eye protection is even more important in today's high-tech living. Researchers in Japan found that people who spend a lot of time at their computers and took 5 milligrams of astaxanthin per day for four weeks reported a 46 percent reduction in eyestrain and higher accommodation amplitude, that is, the ability of the lens to properly focus. A lens that is better able to focus is less likely to get fatigued. Other studies using doses from 4 to 12 milligrams a day found similar improvements to eye fatigue, as well as less eye soreness, less dryness, and less blurred vision (Nakamura et al., 2004; Nitta et al., 2005; Shiratori et al., 2005; Nagaki et al., 2002, 2006). A study showed that people who spend heavy amounts of time on computers recovered from their eye fatigue more quickly when they were pretreated with astaxanthin than when they were not given astaxanthin (Takahashi and Kajita, 2005).

Keeps vision strong

Several studies by researchers in Japan have shown that people who took astaxanthin in doses ranging from 4 to 12 milligrams a day showed improved visual acuity (the ability to see fine detail) and depth perception (Sawaki, 2002; Nakamura, 2004).

Helps keep eyes young

Hard-working tissues like the lens and retina of the eye eventually wear out. Because the lens and retina are two vital tissues affected by the aging process, let's look at what science says about how astaxanthin protects them.

Excessive wear-and-tear of the lens makes it stiff and sticky, so it is less able to change shapes and accommodate the changing light of various visual images. A laboratory study from the Massachusetts College of Pharmacy and Health Sciences in Boston showed that astaxanthin helped protect the lenses of rats' eyes exposed to oxidants (Liao et al., 2009; see also Nakajima et al., 2008; Suzuki et al., 2005; Ohgami et al., 2003; Cort et al., 2010). Another study from the School of Pharmacy of Taipei Medical University in Taiwan showed that astaxanthin protected the lenses of pigs' eyes from oxidation (Wu et al., 2006).

Eye tissue, like brain tissue, is not only hyper-vulnerable to oxidation, but also particularly resistant to cell repair and regeneration. One of the reasons we believe antioxidant protection is important for the eye is that the most sensitive cells of the retina, called the macula, once damaged, are very difficult to repair and regrow. Here's a medical lesson as true for the brain as it is for the eyes: Maintaining health is easier than repairing health, and astaxanthin helps maintain eye health.

ASTAXANTHIN PROTECTS JOINTS AND TENDONS

Many workers and athletes experience discomfort from overuse of their muscles, tendons, and joints, especially following long days at the office or a strenuous workout at the gym. Here are some studies showing how astaxanthin helped in this category. Even though the studies were conducted on small populations, the results are promising.

Supports Workers' Muscles and Tendons

With more and more people spending their days typing on computers, keeping muscles, tendons, and joints working optimally, especially in the hand and wrist, is becoming increasingly important. Researchers studied 20 people who experienced wrist pain after overuse in the workplace and divided them up into two groups: thirteen people received 4 milligrams of natural astaxanthin three times a day, and seven people received a placebo. Those given natural astaxanthin reported a 27 percent reduction in daytime pain after four weeks and a 41 percent reduction after eight weeks. The duration of their daytime pain decreased by 21 percent after four weeks and 36 percent after eight weeks (Nir and Spiller, 2002a). While the results were promising, they were not statistically significant because of the small number of subjects. A larger study is needed to measure significance.

Keeps Strong Athletes Strong

In a 2006 study of thirty-three tennis players, twenty-one of whom received natural astaxanthin and twelve a placebo, researchers found that after twelve weeks the astaxanthin group had less arm soreness and improved grip strength (Spiller et al., 2006).

In a small pilot study, astaxanthin was given to weightlifting athletes to see if it could protect them from delayed onset muscular soreness (DOMS) in a study conducted at the Human Performance Laboratories of the University of Memphis. Nine participants were given a daily dose of 4 mg of astaxanthin for three weeks prior to the weightlifting session and during the twelve-day recovery phase. Five participants were given a placebo. Forty-eight hours after the weightlifting session, the group taking astaxanthin perceived less DOMS soreness than the group receiving the placebo (Fry, 2001).

A Nutrient for All Ages
Every age suffers inflammation. Every age needs more inflammation protection. Young children have a young immune system which is why they get sick so often. Teens work their bodies hard during sports and suffer overuse injuries. Pregnant mothers really overwork their bodies. Young adults, a relative healthy period in life, should be in the "prevent mode" mindset, and seniors suffer ailments from tissues wearing out.

ASTAXANTHIN SUPPORTS A HEALTHY IMMUNE SYSTEM

With the immune system, we emphasize "balance" rather than "boosting," since you want the body's immune system to react according to the body's needs. If the immune system underreacts, germs take over and you get sick. If it overreacts, the immune system gets its signals mixed up and attacks the body's own tissue. Astaxanthin can help the immune system in a few ways.

Works with the Body's Own Soldiers

Circulating throughout your body 24/7 are microscopic soldiers that search out and destroy dangerous toxins, germs, cancer cells, and other things that if allowed to get a foothold within the body will make it sick. These immune system cells go by various names, but a few of the most popular are:

- *Macrophages* (literally "big eaters"). These are white blood cells that gobble up harmful foreign stuff before it can harm the body.

- *Natural killer cells* (NK cells). Also called B-cells and T-cells, these are some of the immune system's most diligent and heavy fighters that are called upon for big wars, such as major illnesses.

Inside the Laboratory of Immune System Researchers

Immune system pioneers Drs. Jean Soon Park and Boon P. Chew at Washington State University discovered that astaxanthin boosts immunity in animals. In an important study, they were able to show that it also increases the health of the immune system in humans. Specifically, they measured "markers of immune health," meaning biochemicals that increase in a person's blood when the immune system is working at its best. In their study, Drs. Park and Chew and their team of immune system researchers examined the astaxanthin effect on healthy females of an average age of 21 (Park et al., 2010). The researchers gave them 2 to 8 milligrams of astaxanthin a day for eight weeks in a double-blind, placebo-controlled study, the gold standard of research. After eight weeks of astaxanthin supplementation, they found an overall improvement in the markers of immune system health: an increased level of NK cells and decreased damage to cellular DNA – an area of the cell most vulnerable to damage when the immune system is weakened. It's interesting that in this study a dose of astaxanthin as low as 2 milligrams a day showed improvement in the immune markers.

ASTAXANTHIN PROMOTES SKIN HEALTH

While astaxanthin's effect on keeping skin healthier and looking younger has long been appreciated, recent Hollywood buzz on natural astaxanthin has increased its popularity. More people are now learning about what dermatologists have long known: natural astaxanthin supports skin health during sun exposure. In 2011, England's second-largest newspaper, *The Daily Mail*, reported that both Academy Award–winning actress Gwyneth Paltrow and supermodel Heidi Klum were using astaxanthin to help beautify their skin. The article, titled "Extended Life Pill: Miracle Supplement Promises to Fight the Signs of Aging," lists these benefits for the skin:

- Lessens wrinkles and improves skin elasticity
- Reduces visible signs of ultraviolet damage.

Treats Skin on the Outside and Inside

In my medical practice, I teach it's not only what you put *onto* the skin to protect it, but also what you put *into* it, and that's where astaxanthin shines. When skin ages, we notice wrinkling, thinning, and sagging because of the collection of sticky stuff that infiltrates collagen, the protein and structural fibers of the skin. This weakens the skin's elasticity and causes it to sag. Since most skin damage is caused by oxidation, it stands to reason that natural astaxanthin, a powerful antioxidant, would be good for skin health. Science agrees.

Astaxanthin Supports Skin During Sun Exposure

Sunburn happens from excess skin oxidation. UV light reddens the skin, and repeated overexposure to sunrays causes photo aging: wrinkled, blotchy, thinned, darkened skin. Since astaxanthin is a powerful antioxidant, it's a natural partner with skin health. To study the astaxanthin effect on sun exposure, scientists tested a group of people to see how much UV light was needed to redden their skin or cause mild sunburn. The group was given 4 milligrams a day of Natural Hawaiian Astaxanthin for two weeks, followed by a repeat of the skin-reddening test. The results were that 4 milligrams of astaxanthin per day increased the amount of time it took for UV radiation to redden the skin, demonstrating astaxanthin's ability to support the structure of the skin during sun exposure (Lorenz, 2002).

Since 1995 there have been many studies on animals showing how astaxanthin can prevent photo-aging from UV exposure. Yet only in the past decade has it been increasingly recognized as a protective factor for the skin of humans.

An Internal Beauty Pill

In 2006 a landmark human clinical study appeared in the journal *Carotenoid Research* (Yamashita et al., 2006). This placebo-controlled research studied 49 healthy women with an average age of 47. The women were divided into two groups – one given placebos and the other supplemented with 4 milligrams a day of Natural Hawaiian Astaxanthin. At the end of six weeks, the women taking it reported so many improvements that Natural Hawaiian Astaxanthin acquired a reputation for being an internal beauty pill. Here are some of the findings:

- In a self-assessment questionnaire, over 50 percent of those taking Natural Hawaiian Astaxanthin reported improvements in skin moisture content, roughness, fine lines and wrinkles, and elasticity.
- Dermatologists using instruments to measure skin health and beauty parameters found improvement in fine lines and wrinkles, elasticity and skin dryness, and moisture levels.
- Before-and-after photos showed visible improvements in fine lines, wrinkles, and elasticity.

This study was especially noteworthy since the conclusions were based not only on the observations of the women taking astaxanthin but also on the dermatologists' examinations.

4

ASTAXANTHIN—MOTHER NATURE'S CHOICE

Over my 40 years in medical practice, I've concluded that the best medicines and nutrients are found in nature. Dr. Mother Nature has the longest track record of producing the medicines and nutrients with the two most important qualities: safety and efficacy. Dr. Mother Nature's medicines and nutrients rarely get pulled off the shelf for being unsafe or not working. And Dr. Mother Nature doesn't continually change her mind with fad supplements that enter the market with a flurry and leave with a fizzle.

One of those natural nutrients is astaxanthin. Dr. Mother Nature has always produced antioxidants to balance the normal oxidation that occurs when living organisms burn fuel for energy. What Dr. Mother Nature hadn't counted on was that the modern living of the 21st century would become an epidemic of excess and chronic oxidation. Much of the world's population is literally rusting away.

CHOOSING THE RIGHT NUTRIENTS FOR OUR BODIES

You may wonder, "Why not just pop a pill?" Not so fast. Remember our two qualities of a good medicine or nutrient? Safe and effective. While prescription drugs can be somewhat effective, many are not safe.

Earlier in this chapter, you learned how the body uses biochemicals within your immune system to protect itself and maintain good health. Health means your body is in balance: the right balance of biochemicals is necessary for health maintenance and the body's natural repair processes. That is, "wear and tear equals repair."

Unlike the naturally occurring antioxidants, which are continuously adjusting their dose according to what's going on in your body, medications such as aspirin, ibuprofen, and the prescription medicine Celebrex (known as NSAIDs, or non-steroidal anti-inflammatory drugs) are taken without knowing what exactly the right amount to do the job is. While NSAIDs may help with arthritis and some cardiovascular problems to block swelling and pain and prevent blood clotting, they can also overreact in the body. Instead of reducing the incidence of heart attack and stroke, they can actually cause these illnesses by over-thinning the blood, resulting in gastrointestinal bleeding or even blood clotting in coronary arteries. This is the reason Vioxx was taken off the market in 2004.

Pharmaceutical NSAIDs can also be hard on the heart by increasing blood clotting, and irritating to the gut. Not only can these drugs increase gastrointestinal bleeding, they can increase gastric acid and reduce the intestinal mucus lining that protects against too much acid, leading to gastrointestinal upsets such as ulcers.

To heal the hurts, people continue to reach out for pharmaceuticals. The problem with many of these over-the-counter and prescription drugs is that they heal one hurt, but cause another. Because of this growing concern over the safety of NSAIDs, many people who want to take charge of their health are learning about lifestyle and dietary improvements and are searching for other options to maintain a healthy balance, such as "nutraceuticals," or dietary supplements. Astaxanthin may be one.

Q: WHAT IS THE BEST DOSE FOR ME?

Surveying the scientific studies, it seems that most people would get the astaxanthin effect they need by ingesting:

4 – 12 mg of Natural Astaxanthin a day

Q: WHY ARE THERE DIFFERENT DOSES FOR DIFFERENT FOLKS?

Bioavailability (how much your intestines absorb of what you eat) varies greatly among individuals. In consultation with your healthcare provider, you will need to experiment to find the right dose to produce the health effect you need. You may be a "high astaxanthin absorber" and need a lower daily dose, or you may be a "low absorber" and need a higher dose. You may be very healthy and just want to keep your immune system in balance and enjoy a bit of preventive medicine. In this case, 4 mg a day may be enough. Or, you may be a metabolic mess and need 12 mg a day.

Take a Tip from Nature

Always take astaxanthin supplements with fats at a snack or meal, or in a gel cap containing oil. Both nature and experiments have shown astaxanthin is better absorbed when partnered with fat – think fatty sockeye salmon.

Q: WHERE CAN I FIND NATURAL ASTAXANTHIN?

In Supplements

You can take astaxanthin in supplements, eat it in food, or both.

46

The most scientifically researched astaxanthin comes from Hawaii. I recommend Natural Hawaiian Astaxanthin because it's safety tested and comes from Haematococcus pluvialis algae. Remember, fish eat algae to get this powerful pink antioxidant.

In Seafood High in Astaxanthin

Sockeye salmon

Sockeye averages 1 mg astaxanthin per ounce of sockeye. So to get 4 mg a day you would need to eat four ounces of sockeye salmon each day, or 1¾ pounds per week. I eat almost that much, but realistically few people will, so they will require a supplement. Other wild salmon species contain lesser amounts of astaxanthin, around one-quarter of that found in sockeye. Usually the redder the fish, the more astaxanthin it contains (Juruman, 1997). Other pink seafood such as lobster, crab, and shrimp have extremely small amounts of astaxanthin.

Remember mother's wisdom: "You are what you eat!" Upgrade that to "You are what the fish you eat eats!"

Fish are as "red" as what they eat

Because the content of astaxanthin in seafood depends on how much that fish eats, there is a wide range of astaxanthin concentration in seafood. For example, the average astaxanthin concentration in wild Alaskan sockeye salmon is around eight times the amount of astaxanthin found in farmed Atlantic salmon. (See graph on next page.)

Synergy Tip

For the best antioxidant effect, enjoy the synergistic effect. Antioxidants work best when eaten with an assortment of other antioxidants, capitalizing on a nutritional principle called synergy, meaning foods eaten together increase the health benefits of each of the separate foods, sort of like 1 + 1 = 3 or 4. This is why your mother told you to eat a variety of fruits and vegetables.

Salmon Species	Astaxanthin Range (mg/kg)	Astaxanthin Average (mg/kg)
Wild sockeye	30-58	40.4
Wild Coho	9-28	13.8
Wild pink	3-7	5.4
Wild chum	1-8	5.6
Wild Chinook king	1-22	8.9
Farmed Atlantic	5-7	5.3
Average		**13.2**

Q: ARE ALL ASTAXANTHINS THE SAME?

No! Natural astaxanthin is much more powerful than synthetic astaxanthin. Here's why: Natural astaxanthin is derived from microalgae. Synthetic astaxanthin is produced from petrochemicals. It seems unnatural to eat supplements made from the same oil you put into your car's engine. Also, natural and synthetic astaxanthin molecules behave differently in the body because they are different molecules. While they may have the same chemical formula, there are three differences:

- The shape of the synthetic astaxanthin molecule is different from its natural counterpart and it has not been tested for safety.

- Natural astaxanthin in nature is always paired with **healthy fatty acids** attached to either one or both ends of the astaxanthin molecule.

- Natural astaxanthin, unlike the synthetic stuff, **contains a variety of other nutrients** that work together like teammates. Dr. Mother Nature discovered this vital health principle of synergy millions of years ago, which is why antioxidants seldom exist by themselves in food.

These biochemical perks make natural astaxanthin behave much more healthfully in the body.

Remember, Dr. Mother Nature has had a much longer history of "manufacturing" the right nutrients in the right proportions than have chemical factories. For example, algae, namely Haematococcus pluvialis, the microalgae from which natural Hawaiian astaxanthin is produced, accumulates other carotenoids as a survival mechanism. These additional carotenoids (beta-caroteneand lutein) work in synergy to make natural astaxanthin more powerful than the synthetic stuff.

Fish farmers have noticed that their fish fed natural astaxanthin are healthier than those fed synthetic astaxanthin, probably because it gets into tissues better than the cheaper synthetic stuff. Next time you're in a fish market, notice how much more pinkish-red the "natural" wild salmon are, compared to the pale-pink farmed ones.

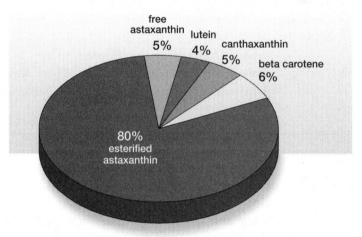

Distribution of naturally-occurring carotenoids in Astaxanthin from Haematacoccus microalgae. Unlike Synthetic Astaxanthin, 80% of the Astaxanthin in the natural, algal-based form is "esterified" with a fatty acid molecule attached. Combined with naturally occurring carotenoids like lutein, beta carotene and canthaxanthin, this esterified Natural Astaxanthin is more potent and effective. It's just what Dr. Mother Nature ordered!

Aquaculture (basically farming in water) experiments to make farmed fish healthier reveal that farmed Atlantic salmon grow better and survive longer when fed natural astaxanthin. As the level of natural astaxanthin in the tissue of baby salmon increased, their survival rates increased from 17 percent to a whopping 87 percent.

Q: WHAT IS A GOOD ANTIOXIDANT DIET?

DR. BILL'S ANTIOXIDANT DIET

Most of a doctor's day is spent treating antioxidant deficiency diseases. Here's the diet I prescribe, which I call the 5 S's:

- **Seafood**: primarily Alaskan salmon, tuna, sardines, and anchovies
- **Smoothies**: multiple dark-colored fruits, berries, organic yogurt, ground flaxseeds, and cinnamon.
- **Salads**: organic arugula, kale, spinach, tomatoes.
- **Spices**: turmeric, black pepper, garlic, rosemary, etc.
- **Supplements** – the three brands that I prescribe are the ones most supported by science:
  — BioAstin, Natural Hawaiian Astaxanthin
  — Omega-3 fish oils, primarily salmon oil from Alaska
  — Juice Plus®, a whole-food-based concentration of many fruits and vegetables

These supplements fill in the gaps when you don't eat 12 ounces of seafood weekly and 9–12 servings of fruits and vegetables daily. And, let's add two more S's: homemade **soups and stews**, which, like smoothies and salads, enjoy the nutritional perk called synergy: Blend many nutrients together and each one becomes more healthful.

TRUSTED TESTIMONIALS FROM HAPPY FOLKS WHO SUPPLEMENT WITH ASTAXANTHIN

Here are some real-life stories from real astaxanthin-supplementers – top scientists, top doctors, top athletes, and everyday folks who just want to feel better and perform better.

PRAISE FROM A TOP PHARMACIST

Pharmacist Suzy Cohen, a nationally syndicated columnist, TV personality, and author of the bestselling books *Drug Muggers, Diabetes Without Drugs,* and *The 24-Hour Pharmacist,* tells me why she takes Hawaiian astaxanthin:

"Why do I take the antioxidant with a funny name? Because Hawaiian Astaxanthin is natural, and it has more cleaning power on the body than vitamin C, E, beta carotene, lutein, zeaxanthin, lycopene, pycnogenol, and alpha lipoic acid. For a good many years, scientists have known that natural astaxanthin helps eye function. But in just the past few years, Japanese researchers have found a fabulous new benefit for the powerful antioxidant. Just 6 milligrams of astaxanthin a day can dramatically improve blood flow to the retina, which nourishes the eyes and helps prevent eyestrain. I take astaxanthin because it helps me with energy, and minor aches and pains. Plus, astaxanthin makes your skin pretty – studies now prove it!"

SELECTED BY A TOP SKIN DOCTOR

Best-selling author and dermatologist **Nicholas Perricone, M.D.**, in his book *The Perricone Promise: Look Younger, Live Longer in Three Easy Steps,* praises the astaxanthin effect on the skin. In his books and as a guest on "The Oprah Winfrey Show," he called astaxanthin a wonderful anti-inflammatory and antioxidant that "gives you that beautiful, healthy glow." He also reported that it reduces age spots and wrinkles. Dr. Perricone attributes astaxanthin's role as an internal beauty supplement to its antioxidant properties of protecting the integrity of skin cell membranes exposed to oxidative stress from sunlight and weather.

NOTED BY A TOP NEUROSCIENTIST

"Astaxanthin has tremendous health benefits. Research on astaxanthin has shown that it reduces markers of systemic inflammation and also reduces damage to DNA. We know from years of research that as we get older there is a background increase in the activity of cells that produce these markers of inflammation. The research links this background of inflammation to the fact that as we age there is a higher chance to develop neurodegenerative diseases and other immune-related diseases. One way to help us to feel younger and have more energy is to take astaxanthin. I have added this to my regimen and feel a lot more energy. There are many hundreds of research studies supporting the benefits of astaxanthin. Astaxanthin is the most potent carotenoid that has been identified and studied." **Paula C. Bickford, Ph.D., Senior Research Career Scientist, James A. Haley Veterans Hospital, Tampa, Florida; Professor, Department of Neurosurgery and Brain Repair, University of South Florida**

TESTIMONIALS FROM TOP ATHLETES

Athletes overuse their muscles, tendons, and joints. "Overuse injuries" account for much of the pain, soreness, and decreased performance that many athletes endure, especially following a marathon performance.

THE TWO A's ARE A PERFECT FIT

Simply put, heavily worked tissues produce excess oxidants and inflammation. Astaxanthin is a powerful antioxidant and anti-inflammatory. **Athletes and Astaxanthin are buddies.**

In my medical practice, I believe that good medicine makes good sense. An athlete's body makes excess oxidants; therefore an athlete needs to eat more antioxidants. You can do that!

Let's see what science says and read the personal testimonies athletes give. While many of these testimonies are from professional athletes, even amateurs, who simply enjoy exercising, may benefit from astaxanthin. In fact, most overuse injuries occur in amateurs, whose muscles are not well trained for a sprint. It's common for these weekend athletes to wake up on Monday morning with sore muscles and joints and realize, "Oh, I overused my body."

"I am fortunate enough to make a living at playing a sport I love and want to continue to do so for as long as I can. I played indoor volleyball for our USA Team and competed in the 1996 and 2000 Olympics. My goal is to play in the 2016 Olympics for Beach Volleyball. I take natural Hawaiian astaxanthin because it's not only one of the best antioxidants, but it's what keeps me playing at my age at a very high level. It's an all-natural immune system builder and helps to give me that extra endurance and strength that I need. It also helps to reduce joint and muscle soreness that is crucial in any athlete's regimen."– *John Hyden, Sherman Oaks, CA*

"I first discovered Hawaiian astaxanthin four years ago when I started racing triathlons, which involves two things: heavy exercise and long-term sun exposure. I was looking for a product that would help my performance in these two areas. Once I started using natural Hawaiian astaxanthin, I noticed a significant improvement in my overuse injuries as well as long-term sun exposure. Antioxidants are the secret to training, performance, and recovery, and natural Hawaiian astaxanthin is a high-quality antioxidant." – *Tim Marr, Honolulu, HI*

"I take natural Hawaiian astaxanthin because balancing life, love, motherhood, and Ironmans, my body endures all kinds of healthy and not so healthy physical stress. Natural Hawaiian astaxanthin is the safest, most effective all-natural form of recovery I've found to help me keep up with it all." – *Bree Wee, Kona, HI*

ACCLAIM FROM JOINT-ACHERS

"I love your product! I was having joint problems in my left arm ... got to the point I couldn't lift a coffee pot or do any pruning at my job without pain. I'm in a physically challenging job as well. I enjoy being able to stay active gardening, working out and snorkeling. Your product blended with glucosamine has really made a difference in my quality of life. The pain in my arm has subsided and my back doesn't bother me nearly as much. I'm now able to do my job without pain. For a while I was very worried that I would have to find a new career and wouldn't be able to participate in my favorite activities." – *Andy Maycen, Kailua Kona, HI*

SKIN HEALTH SUCCESS

"I just got back from my yearly dermatology appointment and got a clean bill of skin health. When she asked how I got so much sun and yet my skin was in such good shape – I attributed it to the high doses of natural Hawaiian astaxanthin. I'm turning 61 this summer and really have very few wrinkles or skin discoloration and yet sport a pretty deep tan from my HOURS of ocean swimming. During this summer season when the sun is so strong, I take more Astaxanthin. I noticed a profound improvement in my skin – many discolorations went away and my elasticity improved. It's a great product and I am so glad you are carrying it in a higher dose. I'm a FAN!!" – **Mariah Dobb, Kamuela, HI**

EXCELLENT FOR ENERGY

"I am a 71-year-old woman living in New York State and I love natural Hawaiian astaxanthin!! When I got Dr. Mercola's newsletter about the benefits of BioAstin, I decided to try it. I have been so pleased that I did. I am an active woman, wanting even more energy, and found that BioAstin supported my body in just that way! More energy was available for all the things I love to do ... garden, yoga, walk, swim, dance! In fact, your product is going on a trip with me (not all supplements do!) on a "Dancing with the Stars" cruise to Alaska!!" – *Deborah Franke Ogg, Olivebridge, NY*

OUTSTANDING FOR OVERALL HEALTH

"I have been taking natural Hawaiian astaxanthin for almost 10 years and have benefited with glowing skin and effervescent energy for my busy lifestyle. Thanks to this reddish wonder, I'm indeed "blushing" with radiant health! Before I discovered Hawaiian astaxanthin, I had a host of recurring health problems including low energy, skin that was dry, and joints that ached for hours after workouts. After just 90 days of supplementation, I began to see improvement in my stamina, skin sensitivity, and joint health – results which have been so consistent over time that I was literally able to throw out 30 other bottles of supplements on my kitchen counter and trade them in for just this one antioxidant superstar." *Susan Smith Jones, Los Angeles, CA*

"I wanted to give a testimonial for natural Hawaiian astaxanthin. I just finished my first order of 4 bottles of the 12 mg format ... and ran out for 2 weeks. I forgot I used to have a sore lower back that would not go away with exercise, massage or other modalities. I noticed big time – as soon as I ran out of your great product – my back has been sore again. I am 56 years old and have been a runner for over 25 years and active otherwise as well. I own my own business – so I need to always be healthy and active. Love it! Love it! Love it! Fan for Life!" – *Debbie Symons, Alberta, Canada*

REFERENCES

Aoi, W., et al. (2003). "Astaxanthin Limits Exercise-Induced Skeletal and Cardiac Muscle Damage in Mice," *Antioxidants & Redox Signaling*, Vol. 5, No. 1: pp. 139–144.

Awamoto, T., et al. (2000)."Inhibition of Low-Density Lipoprotein Oxidation by Astaxanthin," *Journal of Atherosclerosis Thrombosis*, Vol. 7, No. 4, pp. 216–222.

Bagchi, D. (2001). "Oxygen Free Radical Scavenging Abilities of Vitamins C, B, B-Carotene, Pycnogenol, Grape Seed Proanthocyanidin Extract, Astaxanthin and Bioastin In Vitro." Access at www.astaxanthin.org

Beutner, S., et al. (2001). "Quantitative Assessment of Antioxidant Properties of Natural Colorants and Phytochemicals: Carotenoids, Flavonoids, Phenols and Indigoids. The Role of Beta-Carotene in Antioxidant Functions," *Journal of the Science of Food and Agriculture*, Vol. 81, pp. 559–568.

Chan, K. C., et al. (2009). "Antioxidative and Anti-Inflammatory Neuroprotective Effects of Astaxanthin and Canthaxanthin in Nerve Growth Factor Differentiated PC12 Cells," *Journal of Food Science*, Vol. 74, No. 7, pp. H225–H231.

Chang, C. H., et al. (2010). "Astaxanthin Secured Apoptotic Death of PC12 Cells Induced by Beta-Amyloid Peptide 25-35: Its Molecular Action Targets," *Journal of Medicinal Food*, Vol. 13, No. 3, pp. 548–556.

Cort, A., et al. (2010). "Suppressive Effect of Astaxanthin on Retinal Injury Induced by Elevated Intraocular Pressure," *Regulatory Toxicology and Pharmacology*, Vol. 58, No. 1, pp. 121–130,

Curec, G. D., et al. (2010). "Effect of Astaxanthin on Hepatocellular Injury Following Ischemia/Reperfusion," *Toxicology*, Vol. 267, No. 1–3, pp. 147–153.

Deuster, P. A., A. A. Weinstein, A. Sobel, A J. Young (2009). "Warfighter Nutrition: Current Opportunities and Advanced Technologies Report from a Department of Defense Workshop," *Military Medicine*, Vol. 174, No. 7, pp. 671–677.

Fassett, R. G., et al. (2008). "Astaxanthin Versus Placebo on Arterial Stiffness, Oxidative Stress, and Inflammation in Renal transplant Patients: A Randomized, Controlled Trial," *BMC Nephrology*, Vol. 9, p. 17.

Fry, Andrew C. (2001). *Astaxanthin Clinical Trial for Delayed Onset Muscular Soreness*: Report I, Memphis, Tenn.: University of Memphis, Exercise Biochemistry Laboratory.

Grangaud, R. (1951). "Research on Astaxanthin, a New Vitamin A Factor," doctoral thesis at University of Lyon, France. Available on the U.S. Patent and Trademark Office website at www.uspto.gov

Gross, G. J., et al. (2005). "Acute and Chronic Administration of Disodium Disuccinate Astaxanthin Produces Marked Cardioprotection in Dog Hearts," *Molecular and Cellular Biochemistry*, Vol. 272, pp. 221–227.

Hussein, G., et al. (2005). "Anti-Hypertensive and Neuroprotective Effects of Astaxanthin in Experimental Animals," *Biological and Pharmaceutical Bulletin*, Vol. 28, No. 1, pp. 47–52.

Hussein G., et al. (2006). "Anti-Hypertensive Potential and Mechanism of Action of Astaxanthin II: Vascular Reactivity and Hemorheology in Spontaneously Hypertensive Rats," *Biological and Pharmaceutical Bulletin*, Vol. 28, No. 6, pp. 967–971.

Ignarro, Louis J., and Claudio Napoli (2005). "Novel Features of Nitric Oxide, Endothelial Nitric

Oxide Synthase, and Atherosclerosis," *Current Diabetes Reports*, Vol. 5, No. 1, pp. 17-23.

Iwamoto, T., et al. (2000). "Inhibition of Low-Density Lipoprotein Oxidation by Astaxanthin," *Journal of Atherosclerosis Thrombosis*, Vol. 7, No. 44, pp. 216–222.

Izumi-Nagai, K., et al. (2008). "Inhibition of Choroidal Neovascularization with an Anti-Inflammatory Carotenoid Astaxanthin," *Investigative Opthamology & Visual Science*, Vol. 49, No. 4, pp. 1679–1685.

Kim, J. H., et al. (2009). "Astaxanthin Inhibits H202-Mediated Apoptotic Cell Death in Mouse Meural Progenitor Cells via Modulation of P38 and MEK Signaling Pathways," *Journal of Microbiology and Biotechnology*, Vol. 19, No. 11, pp. 1355–1363.

Kim, J. H., et al. (2010). "Astaxanthin Improves Stem Cell Potency via an Increase in the Proliferation of Neural Progenitor Cells," *International Journal of Molecular Sciences*, Vol. 11, No. 12, pp. 5109–5119.

Kudo, Y., R. Nakajima, and N. Matsumoto (2002). "Effects of Astaxanthin on Brain Damages Due to Ischemia," *Carotenoid Science*, Vol. 5, No. 25.

Lauver, D. A., et al. (2005). "Disodium Disuccinate Astaxanthin Attenuates Complement Activation and Reduces Myocardial Injury Following Ischemia/Reperfusion," *Journal of Pharmacology and Experimental Therapeutics*, Vol. 314, pp. 686–692.

Lee, D. H., et al. (2011). "Astaxanthin Protects Against MPTP/MPP+-Induced Mitochondrial Dysfunction and ROS Production in Vivo and in Vitro," *Food and Chemical Toxicology*, Vol. 49, No. 1, pp. 271–280.

Liao J. H., et al. (2009). "Astaxanthin Interacts with Selenite and Attenuates Selenite-Induced Cataractogenesis," *Chemical Research in Toxicology*, February 4.

Liu, X., et al. (2009). "Astaxanthin Inhibits Reactive Oxygen Species-Mediated Cellular Toxicity in Dopaminergic SH-SY5Y Cells via Mitochondria-Targeted Protective Mechanism," *Brain Research*, Vol. 1254, pp. 18–27.

Lorenz, T. (2002). "Clinical Trial Indicates Sun Protection from BioAstin Supplement." Accessed at www.astaxanthin.org

Lu, Y. P., et al. (2010). "Neuroprotective Effect of Astaxanthin on H(2)O(2)-Induced Neurotoxicity in Vitro and on Focal Cerebral Ischemia in Vivo," *Brain Research*, Vol. 1360, pp. 40–48.

Mason, P., et al. (2006). "Refecoxib Increases Susceptibility of Human LDL and Membrane Lipids to Oxidative Damage: A Mechanism of Cardiotoxicity," *Journal of Cardiovascular Pharmacology*, Vol. 47, No. 1, pp. S7–S14.

Massonet, R. (1958). "Research on Astaxanthin's Biochemistry," doctoral thesis at University of Lyon, France. Available on the U.S. Patent and Trademark Office website at www.uspto.gov

Miyawaki, H., et al. (2005). "Effects of Astaxanthin on Human Blood Rheology," *Journal of Clinical Therapeutics & Medicines*, Vol. 21, No. 4, pp. 421–429.

Miyawaki, H., et al. (2008). "Effects of Astaxanthin on Human Blood Rheology," *Journal of Clinical Biochemistry Nutrition*, Vol. 43, No. 2, pp. 9–74.

Monrov-Ruiz, J., et al. (2011). "Astaxanthin-Enriched Diet-Reduces Blood Pressure and Improves Cardiovascular Parameters in Spontaneously Hypertensive Rats," *Pharmacological Research*, Vol. 63, No. 1, pp. 44–50.

Nagaki, Y., et al. (2002). "Effects of Astaxanthin on Accommodation, Critical Flicker Fusion, and Pattern Visual Evoked Potential in Visual Display Terminal Workers," *Journal of Traditional Medicines*, Vol. 19, No. 5, pp. 170–173.

Nagaki, Y., et al. (2005). "The Effect of Astaxanthin on Retinal Capillary Blood Flow in Normal

Volunteers," *Journal of Clinical Therapeutics & Medicines*, Vol. 21, No. 5, pp. 537–542.

Nagaki, Y., et al. (2006). "The Supplementation Effect of Astaxanthin on Accommodation and Asthenopia," *Journal of Clinical Therapeutics & Medicines*, Vol. 22, No. 1, pp. 41–54.

Nakagawa, K., et al. (2011). "Anti-Oxidant Effect of Astaxanthin on Phospholipid Peroxidation in Human Erythrocytes," *The British Journal of Nutrition*, Vol. 105, No. 11, pp. 1563–1571.

Nakajima Y., et al. (2008). "Astaxanthin, a Dietary Carotenoid, Protects Retinal Cells Against Oxidative Stress in-Vitro and in Mice in-Vivo," *Journal of Pharmacology and Pharmacotherapeutics*," Vol. 60, No. 10, pp. 1365–1374.

Nakamura, A., et al. (2004). "Changes in Visual Function Following Peroral Astaxanthin," Japanese *Journal of Clinical Ophthalmology*, Vol. 58, No. 6, pp. 1051–2054.

Nakao, R., et al. (2010). "Effect of Astaxanthin Supplementation on Inflammation and Cardiac Function in BALB/Mice," *Anticancer Research*, Vol. 30, pp. 2721–2725.

Nir, Yael, and Gene Spiller (2002). *BioAstin, a Natural Astaxanthin from Micro-Algae, Helps Relieve Pain and Improves Performance in Patients with Carpal Tunnel Syndrome (CTS)*, Study Report, Los Altos, Calif.: Health Research and Studies Center.

Nishioka, Y., et al. (2011). "The Anti-Anxiety-Like Effect of Astaxanthin Extracted from Paracoccus carotinifaciens," *Biofactors*, Vol. 37, No. 1, pp. 25–30.

Nitta, T., et al. (2005). "Effects of Astaxanthin on Accommodation and Asthenopia-Dose Finding Study in Healthy Volunteers," *Journal of Clinical Therapeutics & Medicines*, Vol. 21, No. 5, pp. 543–556.

Ohgami, K., et al. (2003). "Effects of Astaxanthin on Lipopolysaccharide-Induced Inflammation in Vitro and in Vivo," *Investigative Ophthalmology & Visual Science*, Vol. 44, No. 6, pp. 2694–2701.

Palozza, P., et al. (2009). "Growth-Inhibiting Effects of the Astaxanthin-Rich Haematococcus pluvialis in Human Colon Cancer Cells," *Cancer Letters*, Vol. 283, No. 1, pp. 108–117.

Park, J. S., et al. (2010)."Astaxanthin Decreased Oxidative Stress and Inflammation and Enhanced Immune Response in Humans," *Nutrition and Metabolism*, Vol. 7, p. 18.

Pashkow, F. J., et al. (2008). "Astaxanthin: A Novel Potential Treatment for Oxidative Stress and Inflammation and Cardiovascular Disease," *The American Journal of Cardiology*, Vol. 101, No. 10a, pp. 58d–68d.

Satoh, A., et al. (2009). "Preliminary Clinical Evaluation of Toxicity and Efficacy of a New Astaxanthin-Rich Haematococcus pluvialis Extract," *Journal of Clinical Biochemistry and Nutrition*, Vol. 44, No. 3, pp. 280–284.

Sawaki, K., et al. (2002). "Sports Performance Benefits from Taking Natural Astaxanthin Characterized by Visual Acuity and Muscle Fatigue Improvement in Humans," *Journal of Clinical Therapeutics & Medicines*, Vol. 18, No. 9, pp. 1085–1100.

Sears, W. (2010). Prime-Time Health. *A Proven Plan to Live Happier, Healthier, and Longer*, New York: Little, Brown.

Sears, W., and Sears, J. (2012). *The Omega-3 Effect: Everything You Need to Know About the Supernutrient for Living Longer, Happier, and Healthier*, New York: Little, Brown.

Shen, H., et al. (2009). "Astaxanthin Reduces Ischemic Brain Injury in Adult Rats," *The FASEB Journal*, Vol. 23, No. 6, pp. 1958–1968.

Shimidzu, N., et. al. (1996). "Carotenoids as Singlet Oxygen Quenchers in Marine Organisms," *Fisheries Science*, Vol. 62, No. 1, pp. 134–137.

Shiratori, K., et al. (2005). "Effect of Astaxanthin on Accommodation and Asthenopia-Efficacy-

Identification Study in Healthy Volunteers," *Journal of Clinical Therapeutics & Medicines*, Vol. 21, No. 6, pp. 637–650.

Spiller, G., et al. (2006). "Effect of Daily Use of Natural Astaxanthin on C-Reactive Protein," *Health Research and Studies Center*, Los Altos, CA. Access at www.astaxanthin.org

Sun, Z., et al. (2011). "Protective Actions of Microalgae Against Endogenous and Exogenous Ages in Human Retinal Pigment Epithelial Cells," *Food and Function*, Vol. 2, No. 5, pp. 251–258.

Suzuki, Y., et al. (2006). "Suppressive Effects of Astaxanthin Against Rat Endotoxin-Induced Uveitis by Inhibiting the NF-KappaB Signaling Pathway," *Experimental Eye Research*, Vol. 82, No. 2, pp. 275–281.

Takahashi, N., and M. Kajita (2005). "Effects of Astaxanthin on Accommodative Recovery," *Journal of Clinical Therapeutics & Medicines*, Vol. 21, No. 4, pp. 431–436.

Trimeks Company Study (2003). Access at www.astaxanthin.org

Wang, H. Q., et al. (2010). "Astaxanthin Upregulates Heme Oxygenase-1 Expression Through ERK 1/2 Pathway and Its Protective Effect Against Beta-Amyloid-Induced Cytotoxicity in SH-SY5Y Cells," *Brain Research*, Vol. 1360, pp. 159–167.

Wu, T. H., et al. (2006). "Astaxanthin Protects Against Oxidative Stress and Calcium-Induced Porcine Lens Protein Degradation," *Journal of Agricultural and Food Chemistry*, Vol. 54, No. 6, pp. 2418–2423.

Yamashita, E., et al. (2006). "The Effects of a Dietary Supplement Containing Astaxanthin on Skin Condition," *Carotenoid Science*, Vol. 10, pp. 91–95.

Yazaki, K., et al. (2011). "Supplemental Cellular Protection by a Carotenoid Extends Lifespan via INS/IGF-1 Signaling in C. elegans," *Oxidative Medicine and Cellular Longevity*, Vol. 10, Article ID 596240

Yoshida H., et al. (2010). "Administration of Natural Astaxanthin Increases Serum HDL-Cholesterol and Adiponectin in Subjects with Mild Hyperlipidemia," *Atherosclerosis*, Vol. 209, pp. 520–523.

NOTES

ABOUT THE AUTHOR

William Sears, M.D., is one of America's most trusted physicians. Following his passion for family health and longevity, in this book Dr. Sears reveals a nutrition secret of the sea: Astaxanthin. Together with his wife, Martha, he has written 42 books on parenting, family health, and healthy aging. He has served as a clinical associate professor at the University of Toronto, University of South Carolina, University of Southern California School of Medicine, and University of California, Irvine. Dr. Sears has been on over 100 television shows, including *Oprah Winfrey*, *Dr. Phil*, *20/20*, *Good Morning America*, *CBS This Morning*, *CNN*, the *Today Show*, and *The Doctors*. His AskDrSears.com website is a widely popular health and parenting site, and his contribution to family health was featured on the cover of *Time* magazine in May 2012. Dr. Sears still finds time to actively practice medicine in Dana Point, California, where he lives in a home frequently visited by his eight children and a growing number of grandchildren.